Gunarif Taib
Santosa Wirodiharjo
Helmi Helmi

Development of Small Scale Food Industry Cluster in Indonesia

AF153260

Gunarif Taib
Santosa Wirodiharjo
Helmi Helmi

Development of Small Scale Food Industry Cluster in Indonesia

LAP LAMBERT Academic Publishing

Impressum / Imprint

Bibliografische Information der Deutschen Nationalbibliothek: Die Deutsche Nationalbibliothek verzeichnet diese Publikation in der Deutschen Nationalbibliografie; detaillierte bibliografische Daten sind im Internet über http://dnb.d-nb.de abrufbar.
Alle in diesem Buch genannten Marken und Produktnamen unterliegen warenzeichen-, marken- oder patentrechtlichem Schutz bzw. sind Warenzeichen oder eingetragene Warenzeichen der jeweiligen Inhaber. Die Wiedergabe von Marken, Produktnamen, Gebrauchsnamen, Handelsnamen, Warenbezeichnungen u.s.w. in diesem Werk berechtigt auch ohne besondere Kennzeichnung nicht zu der Annahme, dass solche Namen im Sinne der Warenzeichen- und Markenschutzgesetzgebung als frei zu betrachten wären und daher von jedermann benutzt werden dürften.

Bibliographic information published by the Deutsche Nationalbibliothek: The Deutsche Nationalbibliothek lists this publication in the Deutsche Nationalbibliografie; detailed bibliographic data are available in the Internet at http://dnb.d-nb.de.
Any brand names and product names mentioned in this book are subject to trademark, brand or patent protection and are trademarks or registered trademarks of their respective holders. The use of brand names, product names, common names, trade names, product descriptions etc. even without a particular marking in this work is in no way to be construed to mean that such names may be regarded as unrestricted in respect of trademark and brand protection legislation and could thus be used by anyone.

Coverbild / Cover image: www.ingimage.com

Verlag / Publisher:
LAP LAMBERT Academic Publishing
ist ein Imprint der / is a trademark of
OmniScriptum GmbH & Co. KG
Heinrich-Böcking-Str. 6-8, 66121 Saarbrücken, Deutschland / Germany
Email: info@lap-publishing.com

Herstellung: siehe letzte Seite /
Printed at: see last page
ISBN: 978-3-659-66339-0

CONTENTS

I. INTRODUCTION

Small industries are very strategic role in the development of social economy. This is evident when the economic crisis occurred in Indonesia in 1997. At that time a small industry did not experience any disruption in order to survive. This resulted in many people's economy is not disrupted. But on the contrary, many major industrial production decreased. In addition, a small industry can create jobs in large numbers. By applying the pattern of development and well-planned policy then became the mainstay of small industries in the real sector in Indonesia. Special to West Sumatra, very small-scale enterprises dominate the industrial development. In 2010 small scale industries in number, 93.4 percent and medium scale / large only 6.6 percent.

Especially for small-scale food industries Indonesian government set the direction of development as follows: (1) integrate with the central production of raw materials and other support facilities, (2) in cooperation with large-scale industry, (3) to impose regulations on large-scale food industry (4) develop food industry has high competitiveness.

The Indonesian government through the Ministry of Industry set the Road Map as a reference in the development of industrial clusters. In accordance with the Road Map, the Government of West Sumatra province to form a group that is incorporated in the "Service Unit and Development of Agricultural Products Processing". This business group is the embryo of a cluster of small-scale food industries in West Sumatra. Until the year 2013, has formed 125 groups of small-scale food processing businesses scattered throughout the County / City in West Sumatra.

The formation of industrial clusters in Indonesia are generally stimulated by the availability of raw materials and labor. Usually this industry cluster to cluster in certain areas and are associated with other industries. Linkages to other industries could be a partnership. Interaction between the industry performed on each set of activities ranging from procurement of raw materials, processing comes to marketing. The series of events is a positive synergy in industrial clusters.

Industry cluster has components that work in synergy as a core industry, related industries, suppliers, buyers and supporting institutions. Positive synergy of the various components will determine the level of success of a cluster. In small-scale food industry cluster in West Sumatra, the most influential is the core of industrial components and related institutions. Variables in the core industry is very influential capitalization issues and continuity of supply of raw materials. In supporting institutions that still needs to be improved is coaching related to sanitation production and food security.

Some research on industrial clusters, especially for non-food industrial clusters show that the small industries in the cluster must be able to meet several requirements economically. Research on the ceramic industry Kasongan in Yogyakarta showed that business structure and scale economies dominated by small industry and a partnership with a very good supplier. These conditions resulted in the number of workers who came to the cluster area.

For export-oriented products, the quality of the product to be one of the main requirements. Plered ceramic industry cluster in Purwakarta (West Java), most products are exported, always pay attention to the high quality requirements and continually improve the quality and design. Product quality problems are also encountered in the furniture industry cluster Jepara (Central Java). To produce good quality products required skilled labor in sufficient quantities. If the increase in the number of products is not followed by the availability of skilled labor, the quality of the product will decrease. This resulted in the disruption of cluster developments.

Industrial clusters also influenced by government policies, especially with regard to production costs. Research on coastal fishing cluster shows that the rate of profit is influenced by government policies on fuel pricing. In addition, the low bargaining power in dealing with middlemen resulted fisherman fishermen are recipients of price alone. This is because the middleman is the only fish that direct buyers buy into fishing, so that it can determine the price of fish

The government is also very instrumental in the development of the concept of industrial clusters. Research conducted in the oil industry cluster in East Java showed that this cluster is still in the embryonic stage. For this cluster development is indispensable role of government, especially the ranks of the Ministry of Industry in terms of improvement of cluster organization with a clear membership. It also needs to be designed with the concept of developing a clear vision.

II. SMALL SCALE FOOD INDUSTRY

Small industry has a comparative advantage, although not accompanied by a mastery of technology as a basis for the development of technology-based region. When viewed from the nature and shape, the small industry has the following characteristics: (1) based on local resources so as to utilize the maximum potential and strengthening the independence, (2) is owned and implemented by local communities so as to develop human resources, (3) apply local technologies that can be implemented and developed by local workers, and (4) are scattered in large numbers so that an effective means of equitable development.

Small industry has limited market information and results in lower sales because they can not understand the behavior of consumers. Most small businesses operating with a product-oriented to ignore this aspect of the market. However, there are also small-scale businesses that can compete in the market based on market demand chain. Because of the small and medium enterprises make a major contribution to economic growth.

Efforts to develop small-scale food industries in rural areas initially determined by the ability to identify internal factors (strengths and weaknesses) and external factors (opportunities and threats) which is used as a basis for formulating standards activities and determine the success of business activities. Programs that need to be developed include the development of superior commodities and mainstay, increase value-added agricultural products, the expansion of the market, providing transportation, distribution of products and the development of cooperation. Local resources should be used in the food industry is a commodity that is superior. The flagship product is a product that has a good advantage in terms of production, continuity and competitiveness so accepted by society and can attract investors. Flagship product is a product that is reliable in certain areas because of many cultivated by the local community and have a bright market prospect.

Usually the leading commodity has advantages in terms of economic efficiency, land suitability, market opportunities and the public interest. Leading commodity can also be determined based on the potential of these commodities such as production, productivity, comparative advantage and competitive advantage.

Leading commodity can be grouped as follows:

a. Comparative leading commodity.

Commodities produced a particular area because it has the advantage of natural resources, whereas in other areas of the hard commodities produced. The leading commodity can also be processed products whose raw materials are produced that area.

b. Commodities competitive advantage.

The advantages are produced in a more efficient and effective, so that has added value and higher competitiveness.

c. Leading commodity specific.

Specific commodities and manufactured with high innovation that is superior to other commodities

d. Strategic leading commodity.

Commodities are superior because it has a very important position in the social and economic activity of certain communities.

Determination of the leading commodity agricultural sector can be determined partly by seeing several related factors such as external and internal factors. Internal factors for example relating to the availability and state of the land, plant seeds and human resources involved in the sector. While external factors are the kinds of external factors such as climate / weather, regulations relating to agriculture, relevant agencies, and others.

Determination of the leading commodity region can increase farming efficiency and increase trade between regions and between countries. In addition, the determination of the leading commodity will also increase the competitiveness of the agricultural sector, improve production efficiency, improve productivity and sustainability of development. To increase the productivity and value of sales of products, the development of superior commodities must be adapted to the conditions of agro-ecosystem and sustainable programs. All of this must be supported by good regulation anyway. Leading commodity in the agricultural sector, especially food crops are generally grown on a small scale. Industrial processing of food crops are also cultivated on a small scale by utilizing local labor.

Food industry will have high competitiveness when produced in rural areas that have a continuity of supply of raw materials. Rural food industry development opportunities are still open, both in terms of availability of raw materials and processed products from the demand side. Development constraints are: (1) the quality and continuity of raw materials is not guaranteed; (2) the ability of human resources are limited; (3) the technology applied is very simple, resulting in low quality products; (4) cooperation with large-scale food industry is less well implemented.

Human resources involved in the food industry in rural areas are generally derived from lower layers. They have the skills and education levels are low. For it is necessary coaching to improve his abilities. The skills most needed them include processing techniques, how / division of labor within the group, organizing farming activities, problem solving and planning activities.

III. SMALL SCALE CLUSTER OF FOOD INDUSTRY

Indonesia has known the concept of Small Industry Environment and Industry since 1979. Industry grow through planning and government involvement, while smaller industrial environments grow naturally. The concept was originally developed to foster better small industry in terms of cost. In the mid-1990s, a number of departments also developed the concept of cluster-based industrial development. In fact, there is a department which has published an inventory of the number of potential clusters, ie clusters that have high growth potential.

In general industrial clusters evolved through several stages as follows:
1. Embryonic stage. At this stage, a new cluster is formed with the discovery and innovation of a group of businessmen. Investment and operational costs only come from business owners.

2. Stage of growth. At this stage of product marketing has started well developed. Competition makes this business group seeks to enhance its capabilities.

3. The stage of maturation. At this stage the entire production process has been going well. Customer service has become routine in the attempt. At this stage it appears the new entrepreneurs. The cost of production has to be decisive in the competition to get better profit.

4. Phase setback. At this stage there is a change, the production process changes and the resulting product has been replaced by substitute products manufactured at lower costs.

At this stage of setbacks required openness and innovation that industrial cluster can survive and thrive. Openness and innovation can be the starting point of the development of new industries. Technology is never static, but always evolving. He also explained that there are two factors that drive technological change the face of change:

1. The logic of science and technology, where technology is never perfect so it is always changing. New knowledge always brings change and create new technologies.

2. The economic pressures, where technology is a major force in economic growth.
New technologies produce new products related to market expansion activities and improvement of production efficiency. For it is always necessary applied research and development through education and training.

Industrial cluster development cycle resulted in changes in the number and size of the core business. Specialization and supporting business size also changes. These changes are needed to adapt to changes in the market and improving the capacity of the production process through innovation and technology. The decision to make changes to the core business as well as supporting efforts made by economic considerations. With this change, the cluster becomes more economical.

Integration and synergy of the industrial cluster can improve the efficiency of production so as to increase profits. Some important factors which become obstacles in the development of industrial cluster is a network of partnerships, technological innovation, human resources, infrastructure, the presence of large companies, competition, services, access to a lesser market and market information, access to business support services, and access to finance. Various factors that become obstacles determines success or failure of industrial clusters. Because it is necessary for the reconstruction to rearrange the industrial cluster in order to become better.

In Indonesia are generally small-scale food industry cluster can not develop properly so that the value added has not enjoyed by small entrepreneurs. Because it takes the right step for the development of industrial clusters of small-scale food, especially the leading commodity based. For example, clusters of small-scale food industries in West Sumatra are still at the embryonic stage, because it needs to be developed to the stage of growth. It is necessary for the reconstruction of the existing industrial clusters.

"Service Unit and Development of Agricultural Products Processing" as the embryo of competitiveness clusters have high potential for creating collective efficiency in utilizing the availability of local raw material resources, manpower with special skills, rapid dissemination of knowledge, and the upstream-downstream linkages are strong. As the embryo clusters of small-scale food industry, the "Service Unit and Development of Agricultural Products Processing" can be developed in an integrated and sustainable. Some relevant institutions should work together so that the industry cluster can function optimally. For that we need to understand the components of an industrial cluster as shown in the following table.

Table 1. Components of clusters of small-scale food industries in West Sumatra

No	Component Cluster	Related Party
1	Industrial Core	processing industry ("Service Unit and Development of Agricultural Product Processing")
2	Suppliers	Suppliers of raw materials / raw materials (Farmers Group)
3	Industrial / Transportation Services	Support services, banking / cooperatives, mass media (advertising)
4	Buyer	broker, shop
5	Related Industry	packaging industry, supporting materials
6	Organization / Institution Support	Department of Agriculture, Department of Health, Department of Industry and Trade, Goods Quality Supervision Center, Center for Food and Drug Administration

For the development of business groups joined in the "Service Unit and Development of Agricultural Products Processing" as the embryo is necessary to study the role of industrial clusters of each component of the industrial cluster in order to make a positive contribution in the formation of clusters of small-scale food industries. The group acts as a core industry in the structure of industrial clusters. The linkage group as a core industry with the other components determine the success of the formation of industrial clusters. Establishment of "Service Unit and Development of Agricultural Products Processing" is based on the potential of each region, particularly with regard to the availability of raw materials. However, based on research that has been done is still encountered many weaknesses as follows.

1. Raw Material Supply

Establishment of "Service Unit and Development of Agricultural Products Processing" already noticed the availability of local raw materials. However, for some areas still encountered problems as follows:

a. The quality of raw materials.

  For some manufacturers are still constrained by the quality of raw materials. One example is the quality of the corn as raw material processed products. The poor quality of maize resulted in poor quality of the product, making it hard to compete in the wider market

b. Continuity of supply of raw materials

  Some commodities are still not continuous supply. At any given time some manufacturers still constrained by continuity of supply of cassava as raw material for processed products. In addition, continuity of supply of taro is still problematic. This resulted in disruption of the continuity of the production process.

2. Mastery of Technology and Production Equipment

a. Processing Technology

     Mastery of the technology is still not perfect, especially with regard to the production process. Generally "Service Unit and Development of Agricultural Products Processing" production does not use a standard process so that product quality is not uniform. In addition, the size of the product is still not pay attention to consumer tastes.

b. Sanitation

  Generally "Service Unit and Development of Agricultural Products Processing" production does not pay attention to sanitation. This affects the appearance and shelf life of products. Poor sanitation resulted in many contaminants so easily damaged products. Guidance from the relevant authorities is needed to address this issue because it affects the quality of the resulting product.

c. Packaging

Packaging weaknesses include the design and size. The design of the packaging has not been informed about the advantages of the product. Packaging size also varied so that is not in accordance with the wishes of consumers. Packing assistance provided by the government through the relevant agencies have not been fully used.

d. Capacity and quality tools

Most "Service Unit and Development of Agricultural Products Processing" has a capacity of processing tools are not in accordance with the sheer number of production. There is the capacity of the tool is too large compared to the amount of production, and some are otherwise. There is still a group of businesses that use the means of production are of poor quality. This will affect the production process and the quality of the resulting product.

e. Specifications tool

Tool technical specifications do not fit the needs of the production will reduce the efficiency of production, even unused altogether. Especially for vacuum frying equipment, electrical power required is not in accordance with electrical power in rural areas. It is also not in accordance with the raw materials available and the existing market prospects.

3. Human Resources

Human resource deficiencies need to be addressed, especially with regard to the following:

a. Motivation in trying

It is necessary to motivate business groups through training / education in the spirit of entrepreneurship that is shared by all members.

b. Skills

Improved skills through training needs to be done with the material production technology, business management, maintenance and repair tools, sanitary production, marketing and institutional

4. Business Management

a. Production Planning

Member Services Unit and Development of Agricultural Product Processing "must exist which have the ability to make production planning related to raw materials, production processes and quality control. Planning is used as a reference made in the production process and quality control. With the planning and supervision of the evaluation can be done on a regular basis for continuous improvement.

b. Financial administration

Activities should be well documented, particularly in terms of financial administration. It needs to be prepared to understand the power of simple bookkeeping to record all financial transactions group

5. Marketing

All issues related to marketing, either directly or indirectly, must be addressed properly. It is necessary for some of the following:

a. Designing patterns of sustainable procurement of raw materials both in terms of quantity and quality. Thus the continuity of supply can be guaranteed.

b. Pay attention to the type of products that the market demands

c. Need to form a network marketing

d. Share market information among group members

e. The establishment of a professionally managed outlets

f. There must be synergy with farmers as suppliers of raw materials.

So far, cooperation between farmers with "Service Unit and Development of Agricultural Product Processing" is not well ordered and therefore contributes to the continuity of supply of raw materials

IV. COMPONENTS OF FOOD INDUSTRY CLUSTER OF SMALL SCALE

Clusters of small-scale food industry in the form of "Service Unit Development and Processing of Agricultural Products" in West Sumatra consists of several agro-processing enterprises. As a driving force in the cluster set one manufacturer as "champion". The company "champion" this facilitates other businesses in the production and marketing together. Industrial components in a cluster is always associated with all components in the industrial cluster as a core industry, buyers, suppliers, related industries and supporting institutions. Analysis on industrial components needed to see the synergy between components of the industrial cluster. Here are described the component industry in clusters of small-scale food industries in the province of West Sumatra and West Java.

4.1 Core Industries

Highly influential component of the core industries in West Sumatra is "degree of difficulty to get capital" and "level the level of difficulty in obtaining raw materials". The main raw materials on a small scale food industries in West Sumatra supplied by local farmers. Venture capital typically still use personal funds because it is difficult to get loans from financial institutions such as banks and cooperatives.

4.1.1 Raw Materials

Habits of farmers in planting still influenced by social conditions so that production can not be predicted. Under certain conditions the social conditions will affect the business world, including the agricultural sector. Social conditions in West

Sumatra often influence the selection of commodities are planted. This condition affects the continuity of supply of raw materials for the processing industry. Rural food industry development opportunities related to the continuity of supply of raw materials.

In addition to social conditions, the selling price of commodities produced is one of the considerations for farmers in determining the commodities to be planted. Commodity that can be processed into refined products typically increase the selling price due to the added value in the commodity. The added value for the industry to be enjoyed by farmers. Thus, the supply of raw materials to the processing industry to be smooth. The supply of raw materials is greatly influenced the development of industrial clusters.

To overcome this problem the raw materials required cooperation between farmers / farmer groups in the food industry. This cooperation can be facilitated by Farmers Group Association whose members are farmers and owners of the food processing industry. Thus both groups of farmers as producers of raw materials and industrial clusters as a group to be processed in a single institution. These conditions make it easier to cooperate because it is in a container with the same purpose. Research on Agro-Industry Cluster concluded that the government's dominant role in the development of industrial clusters. The government's role is crucial in developing a new cluster. Gradually the role of government will be less necessary in line with the independence that is owned by the industrial cluster. Because the government through its agents in the district and sub-district level is expected to bridge the needs of cluster members like this raw material procurement problems.

4.1.2 Venture Capital

In general, small-scale food industries in West Sumatra has never been associated with banking. It is because the Bank usually only lend to customers who have collateral. Owners of small-scale food industry are usually not able to provide collateral requested by the Bank. These conditions resulted in small-scale food industries have difficulty to increase production.

Availability of capital is one of the main problems in the development of industrial clusters in Indonesia. To overcome this problem required microfinance institutions, facilitated by the government. It specifically is needed in the development of small-scale enterprises. In small-scale food industry cluster, difficulty obtaining capital influence on business continuity. Small entrepreneurs generally have limitations in capitalization, operating costs are usually only available for the production of only a few times. Because it is often these small businesses disrupted business continuity in the event of arrears of the customer. This condition becomes more difficult because usually small businessmen is also weak in terms of the management and financial planning.

To overcome this takes the role of the government to tackle this problem capitalization. The government should facilitate small entrepreneurs to get help capitalization. The most appropriate role undertaken by the government to develop small-scale enterprises is facilitating the formation of net working. It can facilitate the relationship of small businesses with a variety of parties, including access to resources capitalization.

Net working can also be used to smooth the marketing so that small-scale entrepreneurs to market their products well. Payments for products that have been sold can be smooth. This cooperation should create equality between producers and customers. Small entrepreneurs as producers obtain marketing collateral and the smooth financial transactions, and customers also gain certainty will get a supply of products from small-scale businesses.

To bring about equality in net working, the necessary seriousness of the manufacturer and the customer. In this case the government can play an active role, as both sides have a dependency on the government. Without government intervention is usually the owner of the shop which usually market the products of small businesses easily slow payments for products that have been sold. In this case the small entrepreneur has an inferior position.

In certain circumstances there are times when the shopkeepers who have difficulties, especially in terms of continuity of supply. Therefore the government should play an active role in bridging the working relationship between small businesses with other related parties. The success of the government to facilitate net working in small industrial clusters has been performing well on Cluster Fruit in West Java.

Fruit clusters in West Java has well developed so as to attract a number of businesses outside of the cluster so that incoming conduct business cooperation. Thus the cluster is able to form a partnership with many business partners in production. Cooperation between business units with all members of the cluster is progressing well. In addition to having good cooperation with some of the related industries, cooperation among cluster members have also started running well. Department of Industry and Commerce of West Java province in 2012 conducted a survey to see cooperation among the members of these clusters. The shape and intensity of cooperation can be seen in the following table.

Table 2. Intensity Cluster Member Fellow Cooperation in West Java

No	Form of Cooperation Fellow Cluster Member	Percentage
1	Sharing tools	
	- Never	88,89
	- Sometimes	11,11
	- Often	-
2	Share fixtures	
	- Never	95,82
	- Sometimes	-
	- Often	4,18
3	Loan worker	

	- Never	100
	- Sometimes	-
	- Often	-
4	Contract work	
	- Never	38,89
	- Sometimes	22,22
	- Often	38,89
5	Advertising by mouth	
	- Never	70,65
	- Sometimes	29,35
	- Often	-
6	The exchange of information and experience	
	- Never	27,78
	- Sometimes	44,44
	- Often	27,78
7	Learning together	
	- Never	77,28
	- Sometimes	22,22
	- Often	-
8	Unification vision	
	- Never	72,23
	- Sometimes	27,77

Source: Department of Industry and Commerce of West Java Province (2012)

In general, Fruit Cluster members in West Java already has a good market. Marketing is done with regular customers and some are done by the customer is not fixed. Marketing smoothness is partly due to the continuity of production and product quality has been good. Good and smooth production is also supported by the availability of capital. In this case the cluster members are facilitated to access capitalization.

Most governments in developing countries acknowledge that the core industries of the industrial cluster is able to be a driving force for the local business community. The existence of the core industry has also become a reference for industries that produce similar products. Companies that act as a "champion" in the core industry can be a locomotive for other cluster members. Thus, the core industry can be increased in terms of quantity and quality of production.

Industry "champion" has been able to develop well, but still not able to facilitate other industries in the region to improve the quality and quantity of production. So also with marketing, industry acting as a "champion" already has a good market, even routinely market its products outside the region. However, the marketing is done individually and have no impact on other industries which are in the region.

Mastery of technology by the company "champion" has been good, even has the ability to innovate. However, other industries are still in production with poor technology. This situation shows that the core industry is still not able to contribute to improving the quality of the industry together. The movement of individual core industries still alone, although the amount of effort has been much. In this case happens a grouping of new small industries in one area alone.

4.2 Buyer

Highly influential component of buyers in most small-scale food industry in Sumatra Baratt is "customer satisfaction". These conditions resulted in the production of small-scale food industry is unable to compete with food products of large-scale industry. The most appropriate strategy to attract buyers to use / buy a product is to pay attention to his wishes. In this case the purchaser or consumer satisfaction must be considered by the manufacturer. Further described that there are two main things related to consumer conditions, namely:

a. Consumer identity.

In this case the consumer could be the use of direct (user), traders diluent, and institutions. For small-scale food industry manufacturers of consumer identity that they face are the direct users (user). For that we need an understanding of consumer tastes, especially with regard to the nature of the product organoleptic (taste, aroma and physical appearance).

b. demographic factors

In this case to consider the level of income and population. The level of income will affect consumer tastes. It required foresight manufacturers in product quality by adjusting the level of income or purchasing power of consumers.

To enhance marketing efforts are needed most out of every businessman in the face of threats from other manufacturers. Threats can come from the existence of substitute products, it could be of similar products from other manufacturers. Small businesses should be able to face the competition, whether in relation to the bargaining position of suppliers of raw materials, and in the face of buyers with a better bargaining position. Conditions of competition in relation to threats and bargaining power can be seen in the image below:

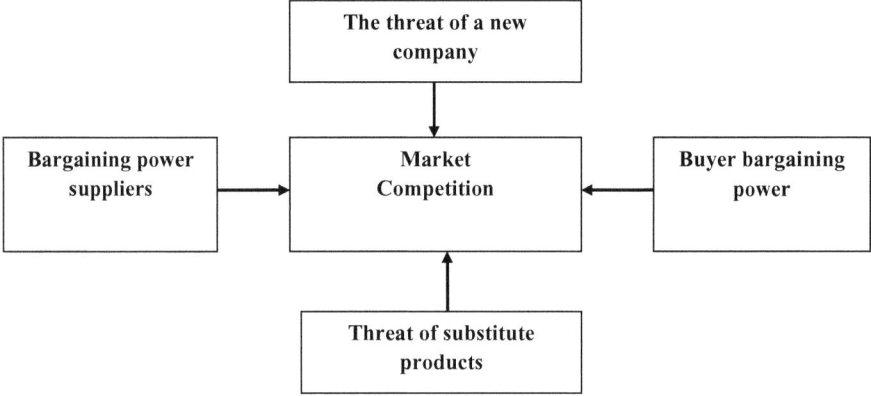

Figure 1. Relationships Product Availability and Bargaining in Business Competition (Purwanto, 2012)

Business competition will always be experienced by all manufacturers. Businesses that already have a clear market need to design its market development.

Measures to be undertaken is the market penetration and product development. In marketing, an increasingly dynamic and competitive indispensable in understanding the ability of manufacturers of consumer tastes. Inability in terms of understanding consumer tastes is the beginning of a marketing failure. Thus, the fulfillment of consumer tastes become indispensable. One disadvantage of small-scale food industry is to meet consumer tastes. Because it is necessary capabilities in terms of satisfying the consumer's taste.

The important thing to consider in marketing: analysis of consumer behavior, competitor behavior analysis and analysis of the general environment. Because it is in marketing, cluster actors already have to understand how to identify competitors, both actual and potential competitors competitors. Steps that must be done is to identify competitors' strategies, goals and objectives of competitors, strengths and weaknesses of competitors and competitors reaction pattern.

To do this competitor analysis is usually small scale industry players are inexperienced. Most small-scale food industry players do not understand their potential. This has resulted in them also have not been able to produce a specific product. Cluster specific needs also vary from one region to another. Industrial cluster should have a specific product. It is necessary for research and development to determine the specific needs of the cluster.

In the development of industrial clusters specific requirements need to be developed so that the product has a comparative advantage. These advantages will increase the competitiveness in the market. For that businesses should always be able to read the market opportunities. The growth of small-scale enterprises are usually heavily influenced by its founder habits and the ability to read the market opportunities.

Good marketing is already owned by the processed fruit products from Fruit Cluster in West Java. Market share is already owned by local consumers, such as fruit processing industry and snacks made from fruit. In general, this fruit cluster members have regular customers in several areas around the fruit processing business location. It also made puree sales contract to other parties such as hotels, tourism and service industries other processing plants. In this case the Department of Industry, West Java continue to act as a facilitator that is beneficial to both parties.

Marketing is done by the members of cluster in a small number still implemented independently. Normally buyers come from local residents who have small-scale processing enterprises such as dodol, jam etc. The selling price agreed upon by the price of a transaction. To help marketing for cluster members that are still in production in small quantities, then the cluster manager will facilitate joint marketing so that it can meet the demand of buyers in large quantities and better prices.

4.3 Supplier

Suppliers are factors that greatly affect the viability of production of the industrial cluster. Supply of goods should be done by a written agreement that both parties have certainty in this business transaction. Farmers who supply through informal channels are less efficient and lower benefit of farmers who supply through formal cooperative. For that we need to be enhanced to the supply of goods in writing with a clear agreement.

Supply of goods, especially raw materials in West Sumatra has not done well. Supply has been done by local farmers, but it is only done individually without involving the group. It is difficult for the industry because there is no certainty of supply of raw materials. In addition to the number of uncertain supply, quality and continuity of supply is not guaranteed.

Incorporated in the core industrial clusters of small-scale food industries in West Sumatra basically be in the Joint Farmers organization. Likewise, farmers who supply raw materials are also members of farmer groups, in which the farmer groups are also in the same group with the food industry. Individual activities in a container the same organization showed that institutional function have not run properly. Under certain conditions, where the supply of the local area numbers are still lacking, the processing industry has been forced to seek supplies to other areas. Of course this requires that transportation costs are relatively more expensive.

Fruit clusters in West Java obtain supplies of raw materials are also of Joint Farmers who are in the area. The main raw materials are supplied from the Joint Farmers are mango, soursop, guava, lemon and strawberry. For the mango supply is not regular because it depends on the season. Joint Farmers usually capable of supplying in October to December each year. Agreed price valid for one year, in the following year made the renegotiation to establish changes in the selling price of fresh fruit to the cluster members.

For a smooth transaction between cluster members with suppliers, then made a contract to purchase raw materials. In this case the "champion" in the fruit clusters act as representatives of the fruit clusters, Joint Farmers Group as a supplier of raw materials. Department of Agriculture Province of West Java in this contract to act as a witness. With such a position that the Department of Agriculture to provide guidance to the fruit growers who are members of the Association of Farmers Group. This assistance is directed so that farmers are able to maintain the quality and continuity of supply to the fruit cluster members.

The interaction between cluster members with suppliers have started running well. Some issues related to cooperation been discussed. However, not all things related to the contract can be discussed, even if there are still lacking intensity needed to be enhanced. Forms of cooperation between cluster members with the supplier can be seen in the following table.

Tabel 3. Fruit Shape Cluster Cooperation with Suppliers

N0	Forms of Cooperation with Suppliers	Persentage
1	Exchange of information and experience	
	- Never	27,78
	- Sometimes	27,78
	- Often	44,44
2	Ease of payment systems	
	- Never	44,44
	- Sometimes	22,22
	- Often	33,33
3	Improving the quality of raw materials	
	- Never	95,86
	- Sometimes	-
	- Often	4,14
4	continuity and continuity of supply	
	- Never	--
	- Sometimes	77,78
	- Often	22,22

Source: Department of Industry, West Java (2012)

4.4 Related Industries

In small-scale food industry cluster in West Sumatra, which affects of related industries is the quality and continuity of supply of packaging. Continuity necessary for the smooth production of packaging supplies, because in addition to affect the sustainability of production, consumers are also very concerned about the quality of packaging. Research conducted by Respati (2012) proved that the level of consumer interest in the product packaging design will enhance the buying interest of 51.5 percent. This figure will rise to 55 percent if the variable is associated with the perception of price. The development of consumer tastes will affect the development of packaging design. Lifestyle changes and differences in values that some segments of society also affects the buying interest and interest in packaging design.

Based on the research results Ervianty (2008) packaging design also affects the impression of the quality of products. The elements of packaging design is very influential composition clearly written, easy to read font usage, striking colors and easy to open and carry. Research conducted Muharram (2011) also stated that attractive packaging design, easy to carry and open is one of the main requirements for the consumer.

To maintain continuity of supply packaging to consider synergies with the packaging business. Anatan (2008) in his research proves that in modern business synergies with related industries is needed. Modern business change competition between independent companies into competition between business networking. Therefore related business networking preformance a production such as suppliers of raw materials, packaging and other suppliers must work together well to produce a product and market it to consumers

Not to ensuring continuity for small industrial packaging buy bottled in small quantities. Manufacturer of packaging usually still reluctant to serve the needs of the packaging in small quantities. To address this need equal partnership between small-scale food industry with packaging manufacturers. Business networking through business partnerships based coordination creating equality in the attempt. Thus, the bargaining position of the parties to be balanced. If the partnership has implemented this pattern, the supply of materials as required in the production of packaging can be guaranteed supply because there is equality between the packaging industry requires packaging manufacturers.

In addition, small-scale food industries which are generally located in rural areas resulted in relatively difficult to access the appropriate packing time. The business location must consider several things such as an efficient means of transport, the number of product offerings in the area and the availability of the required factors of production. Because it required intensive communication and positive synergy between small-scale food industry with packaging manufacturers.

Packing quality also remains a problem for small-scale food industries. This affects the appearance of the product and the production costs are higher because there are a lot of damaged packaging. Industries in rural Indonesia are usually difficult to develop small scale. This is partly due to the low quality of the product and the high cost of production. The low quality of the result of a lack of absorption and limitations of information technology. While the high cost of production due to the low efficiency and market dominance by large-scale industry.

Fruit clusters in West Java has cooperation with several industries that have linkages with the core industry. This linkage could be because it uses the same resources, such as the similarity of raw materials, technology, and distribution networks. The existence of these related industry is the development of clusters. This association led to mutual dependence, both in production, and in terms of marketing. Fruit Cluster in West Java has worked with a variety of related industries in several activities. The company exists located in the vicinity of production, but some are located in other areas (outside the cluster).

4.5 Institutional Support

In small-scale food industry cluster in Sumtera West, very supporting institutions need to consider "coaching on marketing and capitalization" and "guidance on sanitary production and food security". Guidance to small businesses related to the production of sanitation and food safety has not been properly enforced. Consumption and Food Safety Center stated that the food safety factor has not been implemented properly. This resulted in the food consumed there are not safe for the health of consumers.

4.5.1 Development of Marketing and capitalization

Clusters of small-scale food industries still need guidance in terms of marketing and capitalization. Because the government agency that deals with marketing and capitalization must play an active role in the cluster. Seed industry cluster is determined by the success in marketing, effective and efficient institutional, human resource quality and availability of adequate infrastructure.

Market mechanisms should be addressed properly so as to encourage business activities. For the following requirements must be met: (1) implemented in a fair market competition and there is no monopoly power, (2) all regions receive the same information, and (3) regulation of performing well. When these three things can be realized, the market mechanism will be able to run well. In this condition all businesses, including small industry can produce well. This is the responsibility of the government to be implemented.

In developing countries, the monopoly power is still widely practiced by large employers. This resulted in an unhealthy market competition. As a result, the market mechanism can not stimulate economic activity and development are not going well. To overcome this problem the government should take the initiative to encourage economic activity and development through the use of development planning mechanism financed by government investment. This government policy can avoid the monopoly market, and small industries can thrive.

Guidance relating to sanitation production and food security is still needed in the development of small-scale food industries. Especially for small-scale food industry required specifications in production as well as choosing the right technology to be developed in his quest. In addition, external factors such as supporting agencies related to agro-processing industries will also affect the business.

Strong industrial clusters with suppliers of materials and are capable of supporting institutions can improve the productivity of the company. The synergy between the companies will improve the condition of the cluster and the business environment. Thus the cluster development with regard to the success of companies and communities that exist in the cluster.

4.5.2 Guidance on Production Sanitation and Food Safety

Small-scale food industries still require guidance from government agencies that deal with production sanitation and food safety. Guidance needs to be done, especially regarding the processing standards relating to sanitary appliance. Generally small industry is still not optimal attention to sanitation. This result has not been good hygiene products and also affect the durability of the product. Food safety is still negligible, especially with regard to the use of additional food ingredients such as artificial sweeteners, preservatives, artificial colorings and others. Department of Health and the Center for Food and Drug Administration needs to further intensify the development of the small industry. Awareness of small entrepreneurs to food additives is still relatively low. For that besides guidance and supervision should also be given training on sanitation production and food security. It is not a difficult thing because it's basically a small industry generally environmentally friendly. Small-scale industry more environmentally friendly because it does not damage the environment in the surrounding air and water as often happens on a large scale industry.

Coaching is done in line with the requirements of the production permit issued by the local Health Department. This is a weakness for the supervision of the production process is rarely done. In addition, small-scale food industry should be required to create and implement production standards. Likewise with the principles of Good Manufacturing Practises (GMP) shall be applied by all the food processing industry.

The government should provide support for the industrial cluster as a positive impact on the development of the whole industry. For that government institutions related to the industry should play an active role. Support can be given in the form of training for trasnformasi technology, because the technology is needed for the improvement of product quality and production efficiency. Small-scale industry should implement labor-intensive technology, but the application of technology must consider the financial aspects of the business. The combination of the right technology and the ability of the fund would speed up the small businesses thrive.

The role of local governments in the West Suamatera already exist, but still weak in the synergy between government institutions. Training for technological transformation and marketing of only implemented by the Department of Agriculture. Sanitary aspects of production and food security are still not handled properly by the relevant agencies. Industrial processing of agricultural products is basically relatively more serious attention to environmental issues. But still require assistance to environmental conditions more attention.

In the development of the fruit clusters in West Java has a lot of getting contributions from a variety of supporting institutions. Institutions that provide these contributions anybody have direct interaction with the core industry and there are indirect. Various government agencies and the private sector contribute to the development of the fruit clusters. Assistance provided by the government, both central government and the local government carried out by the National Development Program. In the regulation stated that the development of the primary sector, secondary and tertiary developed with the cluster approach.

In connection with this it has been a lot of government agencies, including state and private universities which have played an active role in the development of the fruit cluster. Government support can be seen from the establishment of regulations that support the development of enterprises ranging from upstream to downstream. Local Government also gave attention in terms of capitalization. Among other capital assistance funds from the Corporate Social Responsibility (CSR) which is derived from the State-Owned Enterprises. These funds are channeled through the Fruit Cluster Society of West Java.

Facilitated cooperation covers all aspects of production and marketing. For the development of the government's efforts also conduct training on a regular basis, followed by cluster members. Training materials tailored to the needs of cluster members. Management aspects are also addressed on an ongoing basis. In planning the development of an inventory has been carried out some form of cooperation that has been done and that will be implemented. Details of forms of cooperation can be seen in the following table.

Table 4. Forms of cooperation are facilitated by the West Java Government

Forms of Cooperation	Cooperation existence		Specification
	Yes	No	
Production Aspects			
Financing Access	v		Cooperation Aspects Very Very Good Production
Facilities Access Technology Usage	v		
Facilities machining tools	v		
Facility Availability Raw Materials	v		
Upgrading Facilities Production Engineering	v		
Management aspects			
Facilitating Quality Management	v		Require product branding strategy to make it more widely known cluster
Business Management	v		
Registration of Patents Facilitation		v	
Facilitation of product branding		v	
Training Financial Administration	v		
Aspects of Marketing			
Procurement event to introduce the product	v		Cooperation in promoting clusters is quite good, with an exhibition that
Improving Quality Control	v		
The increase in export volume		v	
Facilitation of market search	v		
Facilitation Partnership with exporters	v		
Promotion via the Internet	v		

exhibition	v		is held twice a year
Aspects of Research and Development			
Human resource development and training	v		
Facilitating partnerships with Research Institutes / Universities	v		Good aspects of R & D Cooperation
Provision of business development		v	
Improved quality and production quality	v		

Source: Department of Industry, West Java (2012)

V. TECHNOLOGY AUDIT

In terms of technology, small industry still faced with the limitations to provide appropriate technology and provide significant added value. This leads to productivity and efficiency produki still low. The last few years the business world, including small-scale agro-industry is strongly influenced by technological innovation. Special efforts need to have a very small scale in the production specifications and choose the right technology to be developed in his quest.

Component technology greatly influenced the development of clusters. Several influential technology components are as follows:

1. Humanware

Knowledge, skills and willingness of actors in the production of food processing industries. Also needed innovation to produce new products / new product development.

2. Technoware.

That need to be observed is a tool / machine that is used, especially with regard to efficiency and produkfitasnya in producing a variety of processed food products.

3. Organoware.

Associated with managerial factors in production. In this case must be observed production management. Observation begins with the raw materials management, production management and marketing management.

4. Infoware.

 All production activities must be well documented. In the food processing industry needs to be observed that no such documentation recording the raw materials used, the processing is done until the recording of sales. This information is necessary for sustainable business.

Examples of the contribution of technology to the small industrial components can be seen from the small-scale metal industry following: technoware contribution to the technology component coefficient of 32.5%; humanware contribution of 20%; infoware component contribution by 25%; organoware component contribution by 25%. Small industry development model should lead to a policy of strengthening the development of technology in small industries that are competitive in the global era, with the order of priority in technoware components, orgaware, infoware, and humanware. Thus it is expected that a small industry can sustain the domestic market, and produce a quality product with a touch of technology.

In addition to producing quality products, small-scale food industry is also expected to absorb labor in the countryside. Small-scale industry should implement labor-intensive technology, but the application of technology must consider the financial aspects of the business. The combination of the right technology and the ability of the fund would speed up the small businesses thrive. By applying technology, the more labor intensive labor can be absorbed.

Clusters of small-scale food industries in West Sumatra in general use simple technology equipment. Technological aspects still need improvement so that production efficiency can be achieved. Because it needed accurate information about the application of technology and some aspects that influence it. From the results of studies that have been done on a small scale food industry cluster in West Sumatra is known that each component has a weight of such technologies in the following table:

Table 5. Parameter Value Component Technology

No	Component Technology	Parameter	Value
1	Technoware	Operational Equipment	0.2889
		Quality of Raw Materials	0.2770
		Increased Production Efficiency	0.1894
		Fulfillment Production Capacity	0.1717
		Technology adoption	0.0489
		source of Technology	0.0242
2	Organoware	Environmental management	0.6267
		Marketing management	0.2797
		Production Management	0.0936
3	Infoware	Production records	0.6267
		promotional Products	0.2797
		documentation effort	0.0936
4	Humanware	motivation	0.5022
		Fellow trust Members	0.1883
		knowledge of business	0.1854
		business skills	0.0817
		innovation	0.0425

Source : Taib (2014)

To accelerate the development of small-scale food industry cluster needs to be evaluated technology components which will affect the growth of the cluster. The results of these evaluations are used to plan the development of small-scale food industry cluster in West Sumatra.

5.1. Technoware

Technoware component is the most important operation of the tool. Because it is on a small scale food industries in West Sumatra Operations tool should be a major concern in preparing the development plan. Small-scale food industries in West Sumatra in general using appropriate tools. The tool is easy to operate and easy to maintain and repair. The tool used is a multi-function, such as a drying (dryer) are used for several different commodities. Likewise, other equipment such as a chopper (slicer), used for some of commodity such as potatoes, cassava, taro, yams and so on.

In fruit clusters in West Java labor generally already have the skills and knowledge in terms of fruit processing technology. Workers already understand the tasks that they are responsible individual, particularly in the production. Production efficiency has been implemented properly in accordance with the existing conditions. The production process at the treatment plant is always concerned with standards of good treatment, although there are limited production facilities and infrastructure.

Limitations include the unavailability of large cold storage capacity. This resulted in only manufacturer capable of doing puree storage for one week, because the storage is only performed using the cooling room. Usually after one week puree shipped to the buyer / factory. In cold storage PT Ultra Jaya puree can be stored for 9 months by using a temperature 2 - 4 ^{0}C.

To improve the control of the processing technology of cluster members are given training on Good Manufacturing Practises (GMP). Materials provided include operational and sterilization of instruments, herb formulation engineered products, how to process products etc. With increasing mastery of this technology, cluster members have been able to improve efficiency to reduce production costs.

Champion cluster already has major equipment (pulper) with high capacity so as to produce as much as 400 kg / hour, with a size of 14-20 mesh, but for a large plant such as PT Ultra Jaya size should be more subtle ie 60 mesh. This is because the processing plant puree into the final product in the form of a drink or a meal. Overall Cluster Fruit already many products made from fruits such as mango puree, puree soursop, guava puree, dried candied mango, candied wet, mango juice, mango syrup, mango powder, lunkhead (mango, guava, soursop), candy fruit- fruits and fresh fruit packaging.

In the field of food technology, there are still limitations to provide appropriate technology and provide significant added value. This causes the heavy dependence on technology for the processing of agricultural products. These conditions result in low productivity, efficiency, and revenue food industry players. The development of the food industry is strongly influenced by technological innovation. Especially for small-scale food industry required specifications in production as well as choosing the right technology to be developed in his quest.

Audit the execution of technology aimed at the implementation of the technology component in the production process. In general, all components of the technology has become a concern for members of the Fruit Cluster in West Java. However, there should be increased so that the development of clusters can be better and provide benefits to all members of the cluster.

5.2. Organoware

Organoware very important component in small-scale food industry cluster in West Sumatra is environmental management. Lack of awareness of producers resulting environmental aspects are not well managed. Cleanliness of the work environment and waste management has not been considered. Processing space is not concerned about hygiene and not well ordered. Thus, the resulting product does not have good quality. Waste processing results just stacked in the vicinity of the processing because it has not provided its disposal container. Because of the location of the business is not clean as much junk scattered. Waste pile is a source of contaminants to the product. These conditions resulted in the food safety of their products is still not good. To overcome the problem of environmental management is needed close supervision of government agencies. This is because it has a dual effect. In a sense this resulted in poor quality products, and besides a bad environment will disrupt the people residing in the vicinity of the processing.

In fruit clusters in West Java, environmental management is associated with the management of production on each cluster member. Environmental management arrangement has been started although the results are not satisfactory. This is associated with assistance activities carried out by a team from the Universities companion. Implementation still vary according to the quality of each resource in the cluster members. As is often the case in most of the small industry, environmental hygiene issues have not been a top priority because they give priority to the production process. Because the existing facilitators make environmental management as one of the priorities in the development of this fruit clusters.

Management of waste management has not received serious attention for most of the fruit cluster members. However, because most businesses only use fruits as the main raw material, the waste generated is not too minimbulkan problem because it is easy to handle. Utilization of waste is still undeveloped so that the principle of zero waste is still difficult to implement.

The process of strategic management in an industry affected by the internal environment, external environment dam general environment. Technology as one of the components of the general environment will provide great opportunities to improve results thus affecting the progress of the company. Middleclass consumer enjoys food products manufactured with attention to environmental aspects. Inclusion of environmentally friendly production processes in the packaging will increase sales volumes in the middle and upper market segments. Small-scale food industry generally still apply traditional marketing strategies. Industries that apply traditional marketing strategies are still trying to convince consumers that they need a product that is produced by the manufacturer. This strategy should be changed, where the manufacturer must design marketing in accordance with the needs of consumers in various market segments. Further stated that the growth potential is classified into three types, namely intensive growth, integrative growth and diversification. In the intensive growth of the company is still looking for opportunities in the scale of operations that the company still relies on market segments that filled it. In integrative growth companies have integrated production activities with the development of the existing market. At this stage of diversified manufacturer has been able to enter the market wider than the existing network.

5.3. Infoware

In small-scale food industry cluster in West Sumatra, a very important parameter of infoware is recording production. Most small-scale food industries in West Sumatra have not done recording production. Because of the economic business conditions are not known with certainty. Thus small businesses can not do development. To address this need training in production management. It is also very necessary guidance berkesimabungan order training materials can be applied appropriately to the production process.

In fruit clusters in West Java documentation production activities have been implemented but still recording in general terms. The training is directed at business documentation is lacking. Recording of production has not been well ordered, this has resulted in the number of regular and detailed production can not be documented. Inventory has not been documented, both of which include raw materials, supplementary materials etc. Need for raw materials and supplementary materials can not be predicted accurately. Thus ordering raw materials have not been associated with a material that is still available. Recording in the warehouse, warehouse both raw materials and products produced pemyimpanan have not done well.

Financial records has been started by some members of the cluster but still not well detailed. This resulted in the cluster members is difficult to calculate the economic benefits obtained. Cluster Facilitator already planning training to improve the ability of members of the cluster in terms of financial management.

5.4.Humanware

Motivation is the most important component humanware on clusters of small-scale food industries in West Sumatra. Because it is very difficult to develop clusters. The lack of motivation resulting in production difficult to compete in the market. The production process is only done by habit. The motivation factor can not be developed if it does not have any business managers to develop self-awareness. Because of the need to create special conditions that this motivation can arise. Type of motivation in trying is as follows.

1. Traditional Model of Motivation Theory

   This model is based on the assumption that the worker does not like his job. In this case the leader needs to provide clear directions and instructions, and oversees the implementation of jobs. Giving a high incentive needed to improve motivation in the work.

2. Human Relations Model

   In this model the workers became saturated with work routines. This resulted in a decrease in the motivation to work. In this case the social relationships is needed. Workers in dire need of its existence in the works. For that leaders must be able to create a harmonious working environment and provide an appreciation of the good work.

3. Human Resources Model

This model states that there are two types of workers, who first lazy workers carry out their duties. For that leaders must force him to work by providing incentives. The second type is a worker who enjoys his job. Workers of this type have a high innovation and creativity. Leaders only create conditions conducive to work.

Increased motivation in the work requires leadership that is able to set a good example in the works. The key task of leadership is planning a program, organize, mobilize and supervise. To achieve this much needed ability to motivate subordinates leader that organizes programs can be run properly. Next was that every leader company must have at least the following three basic capabilities:

1. Conceptual Skills

The leader of the company should be able to create concepts, ideas and ideas for the betterment of the company. It is formulated into a plan that will be operationalized in the company he leads. Planning must be made clearly detailed for easy dimplemantasikan by subordinates.

2. Skills Dealing with Others

These skills are necessary to organize all production activities. Smooth operations in every company really needs the synergy of all existing components. That requires good communication in executing any work plans that have been made.

3. Technical Skills

Types of technical skills varies according to the scope of work. Technical skills are usually needed by leaders in the middle to lower level. This is directly related to production operations. Quality of the product is determined by the technical skills. Good products require clear standards in its production operations. Because of the leadership of the handle should be able to make and prepare operational steps that are structured according to existing treatment standards.

Effort required to develop technological innovation in order to increase the competitiveness of the product. Business competition today is determined by the mastery of technology. Good mastery of technology will result in competitive advantage of a business. Therefore management technologies need to be implemented effectively and efficiently in the internal environment.

Products produced from fruit cluster in West Java is still dominated by semi-finished materials (puree). Innovation to develop the downstream industry has not seen, in addition to producing traditional foods like lunkhead, sweets and jams. Several new businesses to diversify their products based on the size of the packaging, packaging appearance and flavor of the product. This product diversification produces 2 to 3 kinds of new products.

One of the main obstacles is the mindset of the cluster members are still more individulistis so difficult to work in teams. Sense of community is still weak, especially for members of the company has been good. Creative effort is needed to create a sense of togetherness in the group. Members who already have pemasaraan better, have a tendency to not want to provide marketing opportunities to other members.

Mutual trust among members still need to be improved so that progress can be achieved together. Needs further examination together with a fixed limit to the rewards that have more capabilities. For the procurement of raw materials may be easier to do together. This is done in order to more easily obtain raw materials at cheaper prices and guaranteed supply, than if bought individually by a small amount. However in difficult production together. Likewise, in terms of marketing, for those who want marketing independently justified, but for those who are disadvantaged in marketing, should be facilitated for marketing together.

Small-scale food industries in West Sumatra marketing is still at the stage of intensive growth so that marketing is still very limited. They were only able to fill the existing local market. Ability to penetrate the market is still there due to various factors such as capitalization, the sustainability of raw materials and the ability to read the changes in consumer tastes.

Clusters of small-scale food industry requires managers who have experience in marketing. This is evident in the fruit industry cluster in West Java. In this cluster of experience and knowledge of the business manager who acts as a "champion" to serve as inspiration for other cluster members. Thus the cluster members can gradually increase the production and marketing.

VI. REGULATION

6.1 Preparation of Regulation

Indonesia has always adjust the regulation of small industries in accordance with a change, either changes in economic conditions or changes in the corporate world. Therefore in 2008 issued regulations on small and middle business. In these regulations is set on the scale of business, the criteria and definitions of small industries. For the province of West Sumatra government institutions have made regulations to support the development of small-scale food industries. Regulation is made in the framework of guidance and supervision of the industry. Research on government regulation relating to the development of small-scale food industry in Sumatra Various prove that public institutions do not have the same contribution in terms of making regulations. Assessment of the regulations made by several government agencies can be seen in the following diagram:

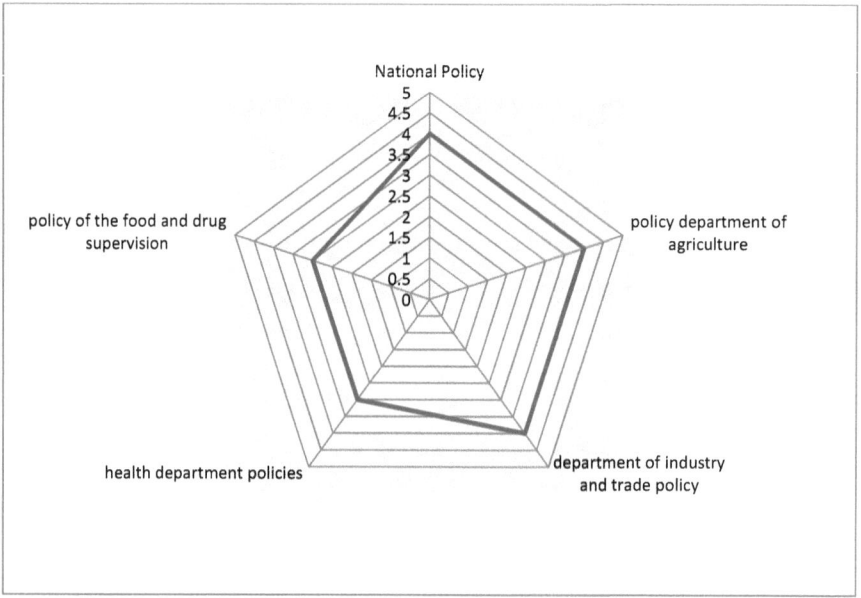

Figure 2. Regulation Related Institutions in the Development of Small Scale Food Industry Cluster in West Sumatra

Specification:
5 = Very Good
4 = Good
3 = Quite
2 = Less
1 = Very Less

6.2 Implementation of the regulation

To perform an analysis of the regulation necessary for the analysis conducted. This was necessary because the results of the analysis is a reference to the revision. Synchronization between the central government regulations and local (district / city) need to be optimized. In an analysis of regulatory or public policy, there are several forms of analysis. Form of policy analysis is as follows:

1. Perspective Model

   This analysis is directed to look at the consequences that occurred before the regulation implemented. This model is also known as predictive models because it uses some forecasting techniques to predict things that will happen as a result of the proposal for a regulation / policy.

2. Retrospective Model

   This analysis is directed to look at the conditions that occur after the regulation is implemented. Also referred to as a model of evaluative as much to evaluate the impact of the implementation of a regulatory / policy.

3. Integrative Model

   This model combines models and retrospective perspective. Also referred to as a model of comprehensive or holistic model for making a thorough study, both before and after the regulation is implemented. In this model use some forecasting techniques and evaluation techniques in an integrated manner. For more details can be seen in the picture below.

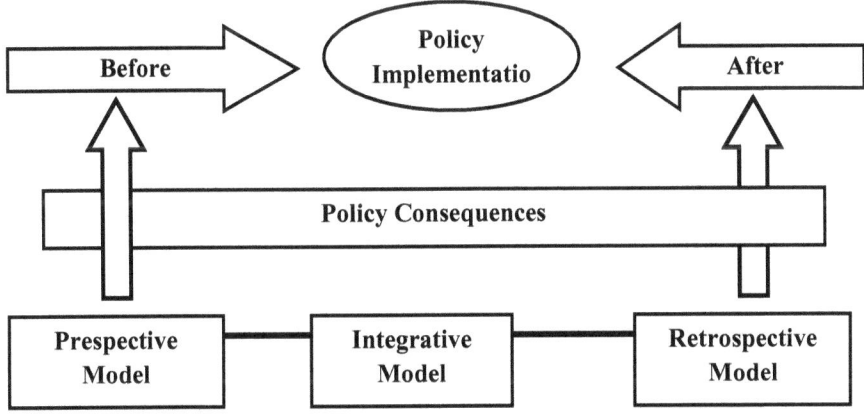

Figure 3. Model for Policy Analysis / regulation (Suharto, 2010)

Very dynamic process of policy analysis as it pertains also to the dynamics of the community in which the policy was applied. The problem to be analyzed needs to be defined, in accordance with its application in society. The problems that arise in the community need to be collected and compiled based on the priority scale. Having collected all the existing problems, then performed a study on the causes of the problem.

The cause of the problem is an important input in evaluating policies. In an evaluation of all positive and negative inventoried. Positive issues sought to be improved implementation. The negative thing studied the causes of, both with regard to the substance of the policy, as well as in its implementation. There are times when there is already a good policy, but weak in implementation. But sometimes also the existing policy does not fit the needs of the local community. In this effort must be found pemasalahnnya roots so that repairs can be done overall yield improvement.

In the analysis performed on the regulation on small-scale food industry cluster in West Sumatra can be concluded that the policy of the central government level, in this case the Ministry of Agriculture and Ministry of Industry, has made regulations for the development of small-scale food industry in the form of clusters. Ministry of Agriculture through the Directorate General of Processing and Marketing of Agriculture established a Strategic Plan which, among others, makes the development of agricultural product processing industry as one of the main program.

The purpose of this program is to increase the added value of agricultural products. For the local level (provincial and district / city) also has created a policy for the development of small-scale food industries. Establish policies that include Agriculture Agribusiness Center Region with the following objectives:

1. Assign each district that has commodity market prospects.

2. Creating the competitiveness of processed products competitive and comparative.

3. Promotion of excellent products out of the region.

4. Guarantee bentu marketing in cooperation with other parties.

5. Increase the value of agricultural products.

6. Application of the concept Agropolitan.

Implementation of existing regulations is still relatively low, which is quite good government agencies in the implementation of its activities is the Department of Agriculture. Lack of implementation of some relevant agencies resulting public response becomes less. This resulted in the development of small-scale food industry cluster in West Sumatra to be slow. It required many efforts to accelerate the development of industrial clusters of small-scale food to increase the welfare of the family farm and reduce poverty.

Policy development of small-scale food industry cluster is appropriate, particularly in terms of location determination. By and large there are six major economic factors that must be considered in determining the location of the business, namely: freight, wage differences between regions, the advantages of agglomeration, inter-regional competition, the concentration of demand and prices and land rent. Based on this, the basis of determining the location of cluster development policy of small-scale food industry can be concluded already very precisely executed.

Regulations relating to the activities of small-scale food industry has been issued by the relevant institutions, both at national and local levels (provincial and district / city). But not everything is as it should be implemented. To determine the extent to which each relevant pihat implementing it, can be seen in the following diagram:

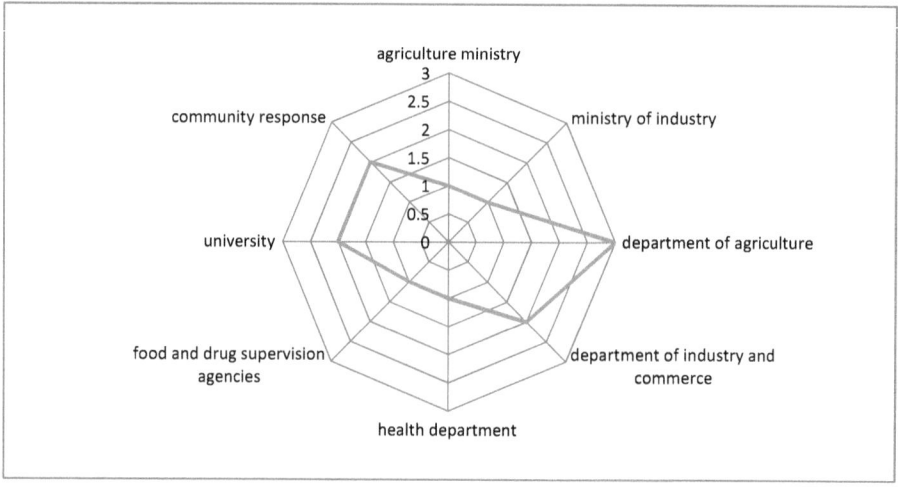

Figure 4. Implementation Regulations Related Institutions in the Development of Small Scale Food Industry in West Sumatra

Specification:
5 = Very Good
4 = Good
3 = Quite
2 = Less
1 = Very Less

To improve speed up the development of small-scale food industry cluster in Indonesia, particularly in West Sumatra, needed a good cooperation between government institutions. Such cooperation is particularly necessary in the case of pembinan and supervision so that clusters of small-scale food industries can develop rapidly and directed.

VII. CONCLUSION

1. Factor Components Industry show that the problem of capitalization and raw materials is still problematic in core industries. Operational capital constraints affecting the sustainability of production so that consumer confidence is reduced. Problems include the raw material quality and continuity of supply of raw materials. The absence of a clear cooperation between industry and farmers / farmer groups resulted in no clarity supply of raw materials.

2. The buyer is very influential factor is the level of customer satisfaction. This problem occurs because the owner of a small industry has not been able to understand trends in consumer tastes. The new production is done based on the ability of production owned, not based on consumer preferences. Consumer tastes have not been a major concern in production.

3. Related Industries influential factor is just the packaging industry. Other related industries has not been a problem because it is a small-scale food industry is not much interaction with other related industries. The main problem that the packaging is packaging quality and continuity of supply. This is partly due to the small industry is still not able to use good packaging because of limited capital owned.

4. Institutional Factors Supporting the role is still to be improved with regard to capitalization and marketing and production sanitation and food safety. Capitalization and marketing plays a very important role in the overall production process. Production sanitation and food safety associated with the quality of the product.

5. Factors relating to the Component Technology humanware factor is still problematic in terms of environmental management. This resulted in the production process has not been well defined, it will ultimately affect the quality of the resulting product.

6. The motivation of business managers still need to be improved. Motivation affects the overall business conditions.

7. Togetherness in core industries are less visible so that activity in the group tends to be done individually only. In a series of operations that began production of raw material procurement, processing until the marketing is difficult implemented jointly.

REFERENCES

Agustiningrum, E.P. 2010. Role Cluster Expectations Batik Pekalongan in Regional Cluster Development SAMPAN (Sapta Mitra Pantura). Thesis at the Graduate School of Engineering Master of Urban and Regional Development. Diponegoro University, Semarang.

Anatan, L., Ellitan, L. 2008. Supply Chain Management. Theory and Applications. Penerbit Alfabeta. Bandung

YN. 2010. The Economics of Small Scale Industries: A look at the Programmes for the Growth and Development of Small Scale Industrie in Nigeria. ICBI 2010, University of Kelaniya Sri Lanka

Bargal, H. 2009. Performance Analysis of Small Scale Industries: A Study of Pre-and Post-liberazilation liberazilation Period. International of Business and Management, Vol 1, No. 2 ISSN 1309-8047

Carlos, JR, Prat Rosario, Rosario Ferriz Marcén. 2006. Influential Factors in Location Choice of Spanish Businesses in Aragon. Journal of Entrepreneurship, Spain. Vol. 15, No. 1, 63-81 (2006)

Clark, J.R., Dwight R. Lee. 2006. Freedom, Entrepreneurship and Economic Progress. Journal of Entrepreneurship, USA, Vol. 15, No. 1, 1-17 (2006).

Dayanto and Abdullah. 2013. Introduction to Management Science and Communication. Penerbit Prestasi Pustaka. Jakarta

Devanath T. 2008. Role of Technological Innovations for Competitiveness and Entrepreneurship. Journal of Entrepreneurship India, Vol. 17, No. 2, 103-115 (2008).

Department of Industry and Trade Cooperative West Sumatra. 2011. Cooperative Industry and Commerce of West Sumatra in Figures 2010. Department of Industry and Commerce of West Sumatra, Padang.

Department of Food Crops of West Sumatra province. 2013. Profile Groups UP3HP in West Sumatra. Department of Food Crops of West Sumatra, Padang.

Ervianty, R.M. 2008. Effect of Packaging Design for Quality (Case Study Shampoo S, XYZ Indonesia). Thesis in Business Management ITB.

Fauzi, H.A. 2007. Analysis of Agricultural Development Strategic Commodity In Depok. Bogor Agricultural University, Bogor.

Govindarajulu., N. 2006. Application of Demand Chain Initiatives to Small Businesses: Journal of Entrepreneurship, Vol. 15, No. 1, 19-35 (2006).

Harisudin, M. 2013. Mapping and Agro-industry development strategy Tempe In Bojonegoro, East Java. Journal of Agricultural Industrial Technology 23 (2); 120-128; 2013

Hendrastuti. 2012. Design of Rural Community Empowerment Model in Agro-Industry Cluster Essential Oil (Case Study: Patchouli Oil) Graduate School of Bogor Agricultural University.

Hilmed. 2003. Development Strategy Agibisnis Commodity Subsector Sawahlunto Plantation in West Sumatra province. Bogor Agricultural University, Bogor.

Hoque, M. J. and Usami. 2008. Effects of Training on Skill Development of Agricultural Extension Workers in Bangladesh: A Case Study in Four Upazilas (sub-district) under Kishoreganj district. Publication: Journal of Social Sciences. , Http://www.accessmylibrary.com. (02 March 2011).

Ishmael, B. 2007. Cluster Development: Policy Conceptualitation and Formulation. The UWI Graduate Institute of International Relations. St. Augustine Campus, Trinidad.

Jayaraman, R., F. Peter Lanjouw. 2004. Small-Scale Industry, Environmental Regulation, and Poverty: The Case of Brazil. The World Bank Review, vol. 18, N0. 3, 2004.

Juzar, Aidil. 2006. Model Cluster Development Strategy Agroindusti Featured Using Core Competencies in the district and institutional. Dissertation at the Graduate School of Bogor Agricultural University.

Kuncoro, M., and Irwan U.S. 2003. Analysis of Cluster Formation Linkage Patterns and Market Orientation: A Case Study Center of Industrial Ceramics in Kasongan,

Bantul, IN Yogyakarta. Empirika Journal Volume 16, N0 1, June 2003.

Majumdar, S. 2008. 2008Modelling Growth Strategy in Small Entrepreneurial Business Organisations. Journal of Entrepreneurship India, Vol. 17, No. 2, 157-168 (2008)

Manik, M.A. 2005. Empowerment Strategy Based Small Industries in Rural Agro-industry. Bung Hatta University, Padang.

Marijan, K. 2005. Developing Small and Medium Industries through the Cluster Approach. Faculty of Social and Political Sciences, University of Airlangga, Surabaya.

Martiningrum, F.D., Sri Gunani and Budisantoso. 2010. Analysis of Industry Cluster Performance Oil and Gas (Oil and Gas) in East Java with System Dynamics Approach. Department of Industrial Engineering Industrial Tenth of November Institute of Technology (ITS), Surabaya.

Muharram, the US, Sofian, S. 2011. Analysis of Product Packaging Design and power Pull Ad for Brand Awareness and Impact on Consumer buying interest. Diponegoro University

Nu'man, A.H. 2008. Small and Medium Industries Development Policy As A Strategy to Increase Competitiveness in the Era of Free Trade. Pustaka Pusat ITB, Bandung.

Partiwi, S., G. 2007. Design of Comprehensive Performance Measurement Model on Cluster System Agroindustrial Sea results. Dissertation at the Graduate School of Bogor Agricultural University.

Porter, M.E. 2011. Creating shared value: Redefining Capitalsm and the Role of the Corporation in Society. Harvard Business School, USA.

Purwanto, I. 2012. Management Strategy. Penerbit CV Yrama Widya Bandung

Respati, N.P. 2012. Effect of Interest in Packaging Design Chocolate Monggo To Buy Consumer Interests (eksplanatif Quantitative Study of the Visitors Center By - By Mirota Batik Yogyakarta Malioboro). Thesis at the University of Atma Jaya Yogyakarta

Rizal, F. 2011. The concept of Fruit Industry Development Network "CIAYUMAJAKUNING" Through the Cluster Approach. Padjadjaran University.

Saikia, H. 2012. Measuring Financial Performance of Small Scale Industry: Some Evidence from India. Journal of Applied Economics and Business Research, 2 (1): 46-54 (2012)

Sandee, H., I. and Sri Sulandari Brahmamntio. 2002. SME Clusters in Indonesia: An Analysis of Growth Dynamics and Employment Conditions. International Labour Office, Jakarta.

Stein K. 2007. Entry Barriers in Rural Business, The Case of Egg Production in Eastern Indonesia. Journal of Entrepreneurship, Norway. Vol. 16, No. 1, 53-76 (2007).

Subkhi, S and Jauhar, M. 2013. Introduction to Theory and Organizational Behavior. Publisher Performance Library. Jakarta

Suharto, E. 2010. Analysis of Public Policy. Penerbit Alfabeta. Bandung.

Sjafrizal. 2008. Regional Economics, Theory and Applications. Penerbit Baduose Media. Padang

Syafruddin, Kairupan, and Limbongan. 2004. Determination System Setup Agriculture and Commodities Based Agroecology Zone in Central Sulawesi. Journal of Agricultural Research, 23 (2), 2004.

Syahza, A. 2003. Economic Analysis of Farm Horticulture As the Regional Agribusiness Commodity Pelalawan Riau Province. Assessment Center for Cooperation and Economic Empowerment (PPKPEM) University of Riau, Pekanbaru.

Taib, G. 2014. Evaluation Component Technology in Small Scale Food Industry Cluster in West Sumatra. International Journal On Advanced Science Engineering Information Technology. Vol. 4 (2014) No. 2

Todd, MS, Bradley JR, John T. 2013. Consumer Valuation of Environmentally Friendly Production Practices in Wines, considering Asymmetric Information and Sensory Effects. Journal of Agricultural Economics, Vol 64 Issue 2, June 2013, page 483-504

Vandeplas, A, Minten, B., Swinnen, J. 2013. Multinationals vs. Cooperatives: The Income and Efficiency Effects of Supply Chain Governance in India. Journal of Agricultural Economics Volume 64, Issue 1 February 2013 Pages 217-244

Wibowo, Yuli. 2011. Design of Industrial Cluster Model progression Sustainable Seaweed. Dissertation at the Graduate School of Bogor Agricultural University.

Wulandari, N.I. 2010. Determination Agribusiness Agricultural Commodities Commodity Based Production Value in Grobogan. Thesis On the Master of Agribusiness, Graduate Program Diponegoro University, Semarang.

Yuhana, S. 2008. Accelerating Development of Agro Commodities Market. Program Development Department of Agro Industry and Commerce of West Java, Bandung.